JUNKERS JU 88

Joachim Stein

One of the first Ju 88 A-1 bombers taken on strength by
Kampfgeschwader (KG) 51.

SCHIFFER MILITARY HISTORY
West Chester, PA

Sources:
Stein Archives
Podzun Archives
Herr Aders, Bonn
Herr Petrick, Berlin
Herr Zucker, Ottobrunn

Translated from the German by David Johnston.

Printed in the United States of America.
ISBN: 0-88740-312-3

This title was originally published under the title, *Das Arbeitspferd der Luftwaffe - Ju 88*, by Podzun-Pallas-Verlag GmbH, 6360 Friedberg 3 (Dorheim). ISBN: 3-7909-0064-8.

We are interested in hearing from authors with book ideas on related topics.

Published by Schiffer Publishing, Ltd.
1469 Morstein Road
West Chester, Pennsylvania 19380
Please write for a free catalog.
This book may be purchased from the publisher.
Please include $2.00 postage.
Try your bookstore first.

Ju 88 A-5 bombers ready for their first test flight at Junkers Bernburg.

Junkers Ju 88

Foreword

The Ju 88 was built in greater numbers than any other German bomber, with over 15,000 being constructed, and is one of the best known of all warplanes.

We are convinced that in Herr Stein, who was involved with the Ju 88 in the repair and replacement parts sector, we have found a competent author. In the following pages he not only supplements existing information, but brings to light several previously unknown facts concerning this aircraft.

Nose of a Ju 88 A-5 with MG/FF cannon installation. The radio technician is compensating the aircraft's radio antenna.

According to the specification issued by the Director General of Equipment (*Generalluftzeugmeister*) of the Luftwaffe in 1935, the Ju 85 and Ju 88 were to be high-speed bombers with a crew of three able to carry 700 to 800 kilograms of bombs. The only defensive armament specified was a single MG 15, as the aircraft were to rely mainly on their speed to avoid enemy fighters. Since Junkers had until now built only very stable and dependable, but slow aircraft, two young men were brought from the United States for the project: their names were W.H. Evers and Alfred Gassner. They began work on 15 January 1936.

The Ju 85 and Ju 88 differed only in their tail surface configurations. The Ju 85 had twin fins and rudders while the Ju 88 featured a single tail unit. Following construction of a mock-up the Ju 85 was dropped. Construction of the Ju 88 V-1 (D-AQEN) began in May 1936. The aircraft's first flight took place on 21 December 1936 with chief pilot Kindermann at the controls. The machine was powered by two Daimler-Benz DB 600 engines of 1,050 h.p. each. The first prototype crashed on one of the early test flights.

The Ju 88 V-2, which differed from the V-1 only in having larger radiators, made its maiden flight on 10 April 1937. The new aircraft proved to be faster than any of its contemporaries. Then followed the Ju 88 V-3 with the new Jumo 211 A engines of 950 h.p. This aircraft achieved speeds of 520 kph over a short course and 504 kph in a longer run. In contrast, the initial versions of the Messer-

Above: The first Ju 88, Ju 88 V-1, D-AQEN, during its maiden flight.

Right: It is a little known fact that the Ju 88 V-1 featured a pressurized cabin.

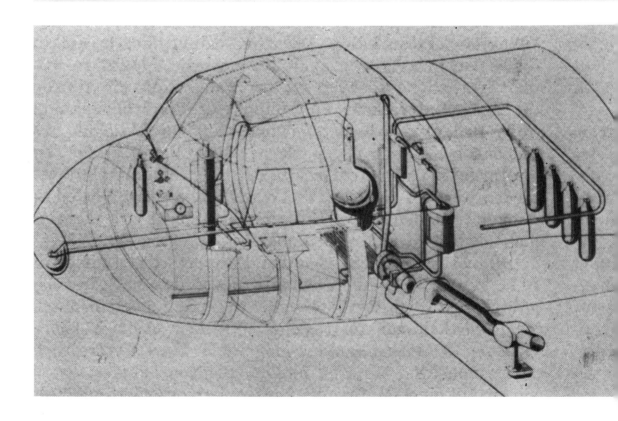

schmitt 109 could reach only 470 kph!

The next prototype was the Ju 88 V-4, which differed from its predecessors in having a nose transparency consisting of flat panels rather than a smooth, rounded nose similar to that of the Do 17 V-1. This aircraft was unable to match the performance of the V-3. Built for propaganda purposes, the Ju 88 V-5 was a modified version built for attempts on various international records. The V-5 made its first flight on 13 April 1938.

Flying the Ju 88 V-5, factory pilots Seibert and Heintz established speed records with 1,000 and 2,000 kg pay loads on 19 March and 30 June 1939, reaching a speed of 517 kph. This high-speed bomber was not built, however, as the Luftwaffe command wanted a conventionally-armed bomber. Four years were to pass before the high-speed bomber concept was finally realized in the Ju 88 S and T. But by then it was already too late.

Incorporating the changes requested by the Luftwaffe command, the Ju 88 V-6 was the prototype for the entire Ju 88 A series. The addition of defensive armament, external bomb racks, new undercarriage and other changes resulted in a reduction in speed of 38 kph while raising takeoff weight. As a result, the Ju 88 A-O, the pre-production series, reached a maximum speed of only 430 kph, or about 100 kph less than the Ju 88 V-3. Dive-bombing capability, which had been demanded by the Luftwaffe, was not achieved. The aircraft was limited to shallow diving attacks. Despite all these shortcomings, which were the result of changes demanded by the Luftwaffe, this aircraft, whose performance was no greater than equivalent British bombers, was the subject of the largest con-

Right: The Ju 88 V-2 shortly after takeoff on its initial flight.

This in-flight photograph of the Ju 88 V-1 shows the elegant lines of the originally planned high-speed bomber version.

struction program in the German aviation industry and was pushed through by General-Director Koppenberg, the head of the Junkers concern, in the face of all difficulties.

A particular source of trouble was the aircraft's hydraulically-activated undercarriage, which had replaced the electrically-activated version used in the Ju 88 V-1 to V-5. The first twenty Ju 88 A-1 bombers, built by Arado in Brandenburg, all suffered undercarriage failures. The Ju 88 was built under licence by Arado, Henschel, the Norddeutschen Dornier Werke in Wismar, ATG in Leipzig and in the Volkswagen factory, while Siebel (Halle), AEG and Opel produced major components. It is interesting to note that from 1943, when the Americans began to bomb the German aviation industry, the installations at VW which were building aircraft parts were heavily bombed while the automobile production facilities remained virtually undamaged.

The first Luftwaffe unit to equip with the Ju 88 A-1 was **I.**/*Kampfgeschwader* (**KG**) **25**, which later formed the basis for **I.**/**KG 30**. The first missions by the Ju 88 A-1 were flown against the British Fleet, with negative results. It was during "*Weserübung*", the invasion of Norway and Denmark, that the Ju 88 scored its first hits on British warships. Only a few Ju 88 bombers saw action during the campaign in the West in 1940.

Above: Nose of the Ju 88 V-6 with the angular glazing specified by the RLM.

Right: The Ju 88 V-6 immediately after its completion, before its registration D-ASCY was applied.

Ju 88 A

From the beginning, three versions of the Ju 88 were planned: bomber, long-range reconnaissance and heavy fighter. An interim *Zerstörer* (heavy fighter) version of the Ju 88 A-1 was built, with three MG 17 machine guns and an MG/FF 20-mm cannon in the standard glazed nose. The A-2 version was planned for operations requiring the heaviest bombs, and differed from the A-1 only in having attachment points for *R-Geräte* (takeoff assist rockets). The A-3 was an unarmed training version with dual controls.

In order to increase the load carrying capability of the Ju 88, it was now fitted with a wing of greater span. The next production version was to have been the Ju 88 A-4; however, the Jumo 211 J engines planned for this series were not available, so the next variant to appear was the Ju 88 A-5 powered by the Jumo 211 G-1. This version, too, received the new long-span wing. A-1 versions still on hand were converted to A-5 standard by adding the new wing and engines. The reconnaissance version of the A-5 was the Ju 88 D-2, which had aerial mapping equipment in the fuselage in place of bomb cells. The D-2 also dispensed with the external bomb racks and underwing dive brakes of the bomber version.

The first production *Zerstörer* version of the Ju 88, the C-2, was developed from the A-5 bomber. Prototypes for the heavy fighter version were the Ju 88 Z-15 and Z-19. The heavy fighter version of the A-1, the C-1, was not built. The C-2 had a faired metal nose with an armament of three MG 17 machine-guns and an MG/FF cannon. Defensive armament consisted of a single MG 15 in the dorsal position. The Ju 88 C-3 was a version

Ju 88 A-5 on a Russian airfield. The aircraft's mainwheels have sunk into the softened earth.

Ju 88 A-1 of KG 51. The short-span wings with ailerons extending to the wingtips are clearly visible.

powered by BMW 801 engines, of which only a single example was built.

The Ju 88 A-5 was first used on a large scale during the Battle of Britain in 1940. During these operations the aircraft's defensive armament of three MG 15 machine guns proved to be inadequate. During the battle this was increased to as many as five MG 15s. At that time the following units were equipped with the Ju 88: KG 30, KG 76 and KG 77 and elements of KG 51, KG 4, KG 40, KG 54, KGr 806 and LG 1 (KGr = *Kampfgruppe*, LG = *Lehrgeschwader*).

The Ju 88 A-6 was developed to deal with British balloon barrages. The aircraft was equipped with a balloon cable fender which extended from one wingtip across the aircraft's nose to the other wingtip and contained cable cutting devices. This installation rendered the aircraft very nose-heavy and was not a success. It was later removed and the aircraft fitted with search radar (*Rostock-Gerät*) as the Ju 88 A-6/U. Like the A-3, the Ju 88 A-7 was a training aircraft. Derived from the A-5, it was equipped with dual controls.

The Ju 88 B series, from which later emerged the Ju 188, was represented only by three prototypes (V 23, V 24, V 25) and a small pre-production batch of ten Ju 88 B-O. Several of these machines flew with KG 40.

A tropical version of the Ju 88 A-1 was planned, but this was dropped in favor of the Ju 88 A-10, a Ju 88 A-5 with tropical equipment. During the Balkan Campaign of 1941 the I. and II. *Gruppen* of LG 1 were equipped with the Ju A-5 and later the A-10. Under commanders Helbig and Kollewe, the two units of LG 1 achieved outstanding success during the battles around Crete and in attacks against English convoys in the Mediterranean. I./LG 1 was especially feared by the English, who knew it by the name of the "Helbig Fliers."

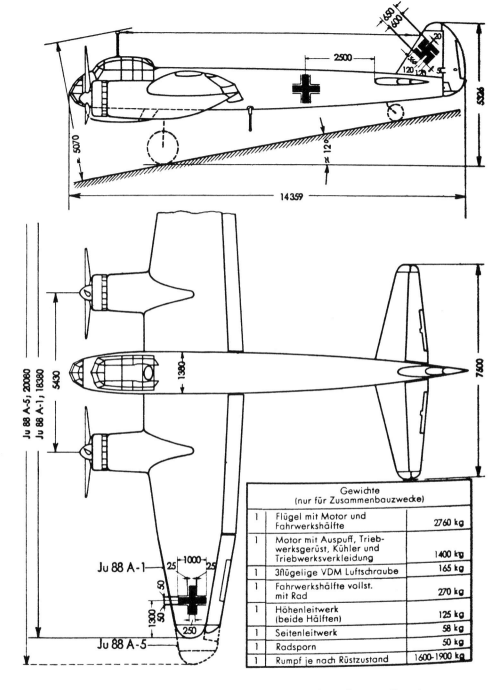

This drawing from the *Flugzeughandbuch* (Aircraft Handbook) T-2088 clearly illustrates the differences between the Ju 88 A-1 and A-5.

Gewichte (nur für Zusammenbauzwecke)		
1	Flügel mit Motor und Fahrwerkshälfte	2760 kg
1	Motor mit Auspuff, Triebwerksgerüst, Kühler und Triebwerksverkleidung	1400 kg
1	3flügelige VDM Luftschraube	165 kg
1	Fahrwerkshälfte vollst. mit Rad	270 kg
1	Höhenleitwerk (beide Hälften)	125 kg
1	Seitenleitwerk	58 kg
1	Radsporn	50 kg
1	Rumpf je nach Rüstzustand	1600-1900 kg

On 21 May 1941 began the transfer of Luftwaffe units to the East. Operation "Barbarossa", the attack on the Soviet Union, was about to get under way. Production of the Ju 88 A-4 had now begun, as the Jumo 211 J, which developed 1,410 h.p. for takeoff, was finally available. The A-4 was to be the most produced version of the Ju 88. While the majority of its predecessors had fabric-covered control surfaces, the Ju 88 A-4 was a true all-metal aircraft. In addition, the aircraft's armor had been strengthened.

The defensive armament of the Ju 88 A-4 originally consisted of five MG 81 machine-guns. This was later increased to one fixed forward-firing MG 81 (*A-Stand*) for the pilot, one flexibly-mounted MG 131 (12.7 mm) for the gunner, two dorsal MG 81 machine-guns (*B-Stand*), and one ventral MG 81Z (*C-Stand*).

With the help of a modification kit it was possible to install a flexibly-mounted MG/FF cannon in the A-Stand position. Other kits consisted of equipment for operation in the tropics or for the carrying of torpedoes. Later these versions were produced in series as the Ju 88 A-11 and A-17 respectively. The Ju 88 A-12 was a training aircraft with dual controls analogous to the A-3 and A-7 versions.

An armored version, the Ju 88 A-13, was built for low-level attacks, which were common in Russia. This version carried fragmentation bomb dispensers or *Waffenbehälter* (WB) 81 (machine-gun pods) on the underwing bomb racks. The WB 81 contained two to eight MG 81 machine-guns, depending on the version in use. The last-mentioned version could fire forwards as well as rearwards.

The Ju 88 A-14 was designed for attacks on sea targets, and featured a fixed, forward-firing MG/FF 20-mm cannon in the forward part of the ventral gondola. These machines

Ceremonial send-off for a Ju 88 unit from its home base.

The first published photographs of the Ju 88 did not depict a bomber as the accompanying caption stated, but the fifth aircraft of the long-range reconnaissance D-2 series. The aircraft's *Werknummer* (05) is plainly visible on the rear fuselage.

The 6. *Staffel* of **KG 30** at Vaernes/Drontheim on 12 June 1941 (Ju 88 A-4).

Ju 88 A-5 landing accident near Jönsvannet/Drontheim in April 1940.

The Ju 88 C-2 *Zerstörer* variant was also employed near Narvik. One of these machines crashed near Björnefell on 7 June 1940.

Remains of a crashed Ju 88 A-4 photographed near Skaiti, northwest of Banak (Norway) in the summer of 1975.

11

Left: Ground personnel of KG 76 refuelling a Ju 88 A-5.

Right: The *C-Stand* with its MG 15 was also the entry hatch for the crew.

Left: This Ju 88 A-5 carries four 50-kg practice bombs on its underwing bomb racks.

Right: In contrast, here live SC 50 and SC 100 bombs lie on the ground ready to be loaded into a Ju 88 A-5.

Above: Maintenance is carried out on the Jumo 211 G of an aircraft of KG 51.

Right: The entry hatch, which also served as the ventral gun position (*C-Stand*), with its lens-type mount for the MG 15, soon proved to offer insufficient protection for the gunner and was replaced by the *Bodenlafette* (Bola) 39 ventral gun mount.

A Look inside the Junkers Ju 88

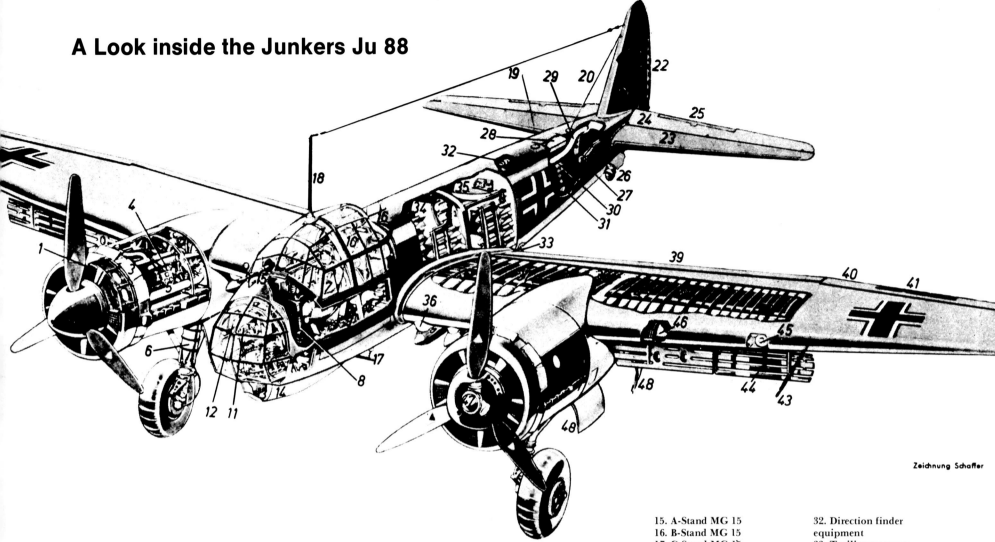

Zeichnung Schaffer

This cutaway drawing provides a good view of the inner construction of the Ju 88 A-1.

14

1. Oil cooler
2. Annular radiator
3. Cooling flaps
4. Jumo 211 engine
5. Indicators
6. Retractable undercarriage
7. Pilot's position
8. Control column
9. Engine controls
10. Jettisonable cabin roof
11. Signal flare box
12. RAB (train release mechanism for bombs)
13. Bomb sight
14. Ventral gondola

15. A-Stand MG 15
16. B-Stand MG 15
17. C-Stand MG 15
18. Antenna mast
19. Antenna
20. Emergency antenna
21. Tail fin
22. Rudder with trim tab
23. Stabilizer
24. Elevator
25. Elevator trim tab
26. Retractable tailwheel
27. First aid kit
28. Inflatable life raft
29. Fuel vent pipe
30. Oxygen bottles for oxygen apparatus
31. Master compass

32. Direction finder equipment
33. Trailing antenna
34. Forward bomb cell
35. Rear bomb cell
36. External bomb rack
37. Fuel tank
38. Oil tank
39. Landing flap
40. Aileron
41. Aileron trim tab
42. Navigation light (port)
43. Pitot tube
44. Dive brake (two-section)
45. Landing light
46. De-icing system
47. Variable-pitch propeller
48. Landing gear doors

The runway had to be cleared of snow before LG 1 could begin winter operations. The aircraft carries two 300 liter fuel tanks on the underwing racks for the ferry flight to the South.

The Mediterranean Theatre

Left: Two Ju 88 A-5 bombers in flight on their way to attack targets on the island of Malta.

Right: KG 30 was also transferred to Sicily for the attacks on Malta. Here a Ju 88 A-5 in Catania. In the background is Mount Etna.

PC 1000 armour-piercing bombs were used in attacks on Malta's stone bunkers.

Under the hot Italian sun ground crews usually worked in bathing suits. Here a Ju 88 is being refuelled.

Captured French Renault tankettes were used to transport bombs to the aircraft.

Left: Ju 88 A-1 of KG 51 before a mission in the West in 1940.

Right: Landing accident by a Ju 88 A-5 during an en route stop in Warsaw.

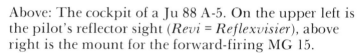

Above: The cockpit of a Ju 88 A-5. On the upper left is the pilot's reflector sight (*Revi = Reflexvisier*), above right is the mount for the forward-firing MG 15.

Above right: Ju 88 A-5 with the new dorsal gun position (two MG 81 machine-guns) shortly before the Allied landings in Sicily.

Right: A Ju 88 A-5 of LG 1 over the Balkans.

Ju 88 A-5 of KG 77 on an Eastern Front airfield. Judging from the worn paint on the propeller spinner, the aircraft has already seen plenty of action.

Left: Days of rain have left this KG 51 airfield under water, making take-offs and landings very difficult.

Right: View from the nose compartment of a Ju 88 A-5 of the aircraft ahead. Condensation trails are visible as the aircraft are flying at high altitude.

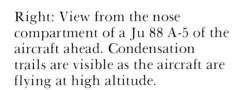

Ju 88 A-5 of LG 1. The ground crew has just made the aircraft ready for flight.

Ju 88 A-4

Left: Ju 88 A-4 of KG 76 over the Soviet Union. The white band around the rear fuselage indicates that the aircraft had been employed in the Mediterranean Theatre a short time before.

Ju 88 A-4 flown by a *Staffelkapitän* of KG 76.

Left: A Ju 88 A-4 of KG 54 which was left behind during the German retreat in North Africa due to lack of fuel and captured by the British.

Right: Ju 88 A-4 on a Sicilian airfield during the attacks against Malta. In the foreground are SC 1000 bombs.

Left: Ju 88 A-5 of LG 1 on an airfield in the West.

Below: A Ju 88 A-5 of KG 76 during an attack on Krasnogvardeysk.

Above left: The Ju 88 D-2 was the reconnaissance version of the A-5. Here is a Ju 88 D-2 of the 4. *Staffel of Aufklärungsgruppe* 14 (5F + DM).

Above: Overly harsh braking resulted in this Ju 88 standing on its nose on the Prague-Ruzyn airfield.

Left: This Ju 88 A-5 was shot down near Krasnogvardeysk on 25 December 1941.

The black-yellow pennant of the *Gruppenkommandeur* flutters from the cockpit of this Ju 88 A-5 in the summer of 1941.

Below: Ju 88 A-4 of KG 51 in the winter of 1941/42 in Russia. The aircraft's engines are being run up, while its bombs are still lying on the ground.

Left: A Ju 88 A-4 of one of the *Kampfgeschwader* deployed in the East. It has the yellow fuselage band worn by aircraft operating over the Eastern Front.

Facing page above: The commander of a Ju 88 D-1 discusses a reconnaissance mission with his crew.

Right: A Ju 88 A-10 of LG 1 on an airfield in North Africa. The aircraft is wearing the sand-color tropical camouflage scheme.

Left: Ju 88 A-4 of KG 76 being serviced.

Right: Compensating the antenna installation of a Ju 88 A-4. A good view of the left dorsal gun position (*B-Stand*) with its MG 81 machine-gun.

Above: A dragon was featured in the *Gruppe* emblem of I./KG 76.

Above right: Ju 88 A-4 (F1 + BR) of KG 76. The aircraft was used in the Mediterranean Theatre, therefore the tropical camouflage scheme. The machine is being refuelled.

Right: In this photograph of a Ju 88 A-4 both dorsal gun positions and their MG 81 machine guns are visible.

Facing page: Ju 88 A-4 with special locating devices for anti-shipping sorties over the Mediterranean, 1943. FuG 200 *"Hohentwiel"* radar mounted in the nose, FuG R *Neptun* above the wings and FuG 101 beneath the wings.

Facing page: The Ju 88 A-4 flew the majority of sorties mounted against English supply convoys destined for the Soviet Union.

Right: Ju 88 A-4 of *Kampfgruppe* 806, formerly *Küstenfliegergruppe* 806, at Catania, Sicily in 1942.

Left: Ju 88 A-4 of an unknown *Kampfgruppe*, probably on the island of Sicily, as the white fuselage band suggests the Mediterranean Theatre.

were committed to the battle in the North Atlantic, sometimes equipped with search radar.

A further version of the A-Series was the A-15. It featured a large, bulged bomb bay beneath the fuselage, which could accommodate the largest bombs. Only one small series of this version was built. The Ju 88 A-16 got no farther than the drawing board, while the Ju 88 A-17, as already mentioned, was a torpedo bomber derived from the A-4. This brought the series of Ju 88 A variants to an end.

The C-2 and C-3 *Zerstörer* versions have already been described. The Ju 88 C-4 was similar to the C-2 with some internal improvements. It was followed by the C-5, which was powered by BMW 801 engines. In contrast to their predecessors, these two versions had the ventral gondola replaced by a small weapons pod containing two fixed, forward-firing MG 17 machine-guns. The offensive armament of the Ju 88 C-5 was thus five MG 17 machine-guns and an MG 151/20 (20-mm) cannon. Defensive armament consisted of a single MG 15 machine-gun. The aircraft's maximum speed was 570 kph. Because of the shortages of the BMW 801 engine, which was needed for fighter (Fw 190) production, plans for series production of the C-5 were abandoned.

The Ju 88 C-6, the first fighter version derived from the Ju 88 A-4, began to leave the production lines in the autumn of 1942. Once the first airborne radar, the *Lichtenstein* C-1, became fit for operational use, these aircraft were used almost exclusively by the night fighter units. One exception was the 1. *Staffel* of *Nachtjagdgeschwader* 3, which was deployed to North Africa. It was equipped

Above: The nose of a Ju 88 C-6 of NJG 1 with *Lichtenstein* radar array.

Right: Weapons installation in the nose of a Ju 88 C.

with Ju 88 C-4 and C-6 fighters without airborne radar.

The offensive armament of the Ju 88 C-6 was very powerful, consisting of three MG 17 machine-guns and three MG/FF cannon. Despite the loss in speed caused by the antenna array of the *Lichtenstein* radar, the C-6 attained a maximum speed of 550 kph. The aircraft's effectiveness as a night fighter was later enhanced with the adoption of the *Lichtenstein* SN 2 radar, which was designated FuG 220.

Just as the Ju 88 D-2 reconnaissance aircraft was derived from the A-5 bomber, the Ju 88 D-1 was a reconnaissance version of the A-4, with a range of nearly 3,000 km. The later Ju 88 D-5, also a derivative of the A-4, possessed a range of 3,100 kilometers using external fuel tanks. The E and F versions of the Ju 88 got no farther than the mock-up stage. From them were developed the Ju 188 E and F.

The superiority demonstrated by Soviet tank units resulted in attempts to destroy the Soviet tanks from the air. Rudel's success with the Ju 87 G is well known; however, the Ju 88 was also employed against enemy armor. The first trials were undertaken with a Ju 88 A-4. The bomber's ventral gondola was removed and replaced by a larger one containing a 75mm Pak 40 with muzzle recoil brake. It was found, however, that the jet of exhaust gases struck the propeller blades, and the trials had to be broken off. Attempts were now made to employ the Ju 88 P-2 and P-3, which were armed with two 37mm cannon, against American bomber formations. However, these trials were also unsuccessful.

Luftwaffe airfield crews are equipped with the latest foam fire-extinguishing equipment to combat engine and fuel fires.

Above: This Ju 88, code Fl + GM (Werknr. 4339), belonged to KG 76 and carried the white fuselage band of the Mediterranean Theatre.

Above left: A Luftwaffe ground crew uses a Renault tankette to haul an SC 1800 (1,800-kg) high-explosive bomb to a Ju 88 A-4 of *Lehrgeschwader* 1.

Left: A *Kette* of Ju 88 A-4 bombers of KG 1 on a mission over Russia. All three exhibit the yellow fuselage band worn by Luftwaffe aircraft on the Eastern Front.

Left: Refuelling a Ju 88 D-1 long-range reconnaissance aircraft of *Aufklärungsgruppe* 122.

Below: A field conversion by KG 51 of a Ju 88 A-4 for the *Zerstörer* (heavy fighter) role. The aircraft carries an additional armament of four MG 81 machine guns.

Tests were then carried out with the Ju 88 P-5, which was armed with the 50mm KwK 39, in an attempt to shoot down American B-17 and B-24 bombers attacking by day. However, the machine proved to lack the necessary maneuverability. The aircraft was then given to the night-fighters, who rejected it.

Somewhat different were the long-range Ju 88 H-1 reconnaissance aircraft and H-2 fighter. Production of the latter was limited to one prototype. The H-Series aircraft were all converted from Ju 88 D-1 reconnaissance aircraft or C-6 fighters. Lengthening the fuselage to 17.65 meters permitted the inclusion of additional fuel tanks. The aircraft were powered by BMW 801 D-2 engines. Equipped with FuG 200 *Hohentwiel* search radar, the Ju 88 H-1 proved successful as a long-range reconnaissance aircraft in the Battle of the North Atlantic.

Ju 88 G-1

Left: The crew of this Ju 88 G-1 deserted to England with their aircraft, which was the subject of extensive tests carried out by the Royal Air Force.

Right: This Ju 88 G-1 fell into British hands later. In the background are several Spitfires.

Right: This Ju 88 A-4 (Werknr. 740) of the *Erprobungsstelle* (Test Establishment) Gotenhafen suffered an undercarriage failure on takeoff while carrying an L 10. Further trials were conducted using the He 111 H-6.

It has already been mentioned that the concept of the *Schnellbomber*, or high-speed bomber, was revived toward the end of the war. A standard Ju 88 A-4 was converted in 1943/44. It received a new fuselage nose, not quite as slim as that of the Ju 88 V 3, but similar, and the ventral gondola and under-wing racks were dispensed with. The aircraft's only defensive armament was a single MG 131 machine-gun. Power plants were two BMW 801 G radial engines, and the aircraft could carry 800 kilograms of bombs internally. The machine received the designation Ju 88 S-O. A small S-1 series was built and used in action. The Luftwaffe now had a high-speed bomber capable of 600 kph.

A long-range reconnaissance version, designated the Ju 88 T-1, was also built. Planned further developments, the S/T-2 and -3, did not reach production. The first unit equipped with the Ju 88 T-1 was 2.(F)/123.

In mid-1943 improved versions of the Ju 88 C-6 began to reach the night fighter units. The new Ju 88 R-2 was basically a C-6 with BMW 801 D-2 engines, equipped with FuG 202 and FuG 212 radar. The aircraft was not exactly easy to fly, as the tail surfaces of the normal C-6/A-4 were too small for the more powerful engines. Nevertheless, good results were obtained with the type. Nevertheless, the R-Series was only an interim solution pending introduction of the redesigned Ju 88 G-1 night fighter. This aircraft was also powered by BMW 801 engines, but featured the larger tail surfaces of the Ju 188. The large ventral gondola disappeared. In its place was a flat fairing housing four forward-firing MG 151/20 cannon. Two further MG 151/20 cannon were installed in the fuselage nose and an MG 131 machine gun in the dorsal position. Production aircraft were fitted with the FuG 220 *Lichtenstein* SN 2 radar. The G-1 was flown by most German night fighter units.

Above: The Ju 88 GV-1, prototype for the G-1 series. The aircraft was equipped with the *Lichtenstein* C.1 radar as the FuG 202 was not yet available.

Below: Ju 88 G-1 (D9 + NL) of IV./NJG 2 on 6 October 1944 at Bulltofta.

The Ju 88 as Test Bed

Left: The Ju 88 A-4 was also used in aero-engine development. Here it is serving as a flying test bed for one of the first Walter rocket engines.

Right: Also tested on the Ju 88 A-4 was the Junkers Jumo 004 jet engine.

The Ju 88 in Foreign Service

Left: The Finnish Air Force received several Ju 88 A-4 bombers, which, later in the war, were employed against German forces.

Right: Several Ju 88 A-4 bombers were also delivered to the Rumanian Air Force.

The Ju 88 G-2 and G-3 remained projects, while the G-4 was built in small numbers and differed from the G-1 only in the addition of the FuG 227 *Flensburg* homing device. The G-5 version was abandoned before entering production. The Ju 88 G-6 followed in 1944. It was powered by the Jumo 213A of 1,750 h.p. and was otherwise similar to the G-4. A conversion kit was available for the installation of two upwards-firing MG 151/20 cannon, the so-called *"Schräge Musik."*

Only a few examples of the G-7 version were built. It was similar to the G-6 except that the larger wing of the Ju 188 had been adopted for improved climb performance. The G-6 and G-7 versions had an extensive array of radio equipment: FuG 16 ZY, FuG 101a, FuG 220, some aircraft also the FuG 228, and finally the FuG 227.

The tragedy of the Ju 88 lay in the fact that an aircraft with superior performance had been developed which could have become the first true high-speed bomber (in the manner of the later De Havilland Mosquito). However, due to the inability of the men in charge of German air armaments to accept this new concept, it was made into a conventional medium bomber which, by 1943, was already obsolescent compared to American and British designs. As produced, the Ju 88 was, like the He 111, a workhorse, but not a high-performance aircraft.

During and after the war the Ju 88 was flown by foreign air forces. Among Germany's allies, Italy, Rumania and Finland received small numbers of the Ju 88 A-4. Following the war Spain and France flew the Ju 88 A-4 as well as a small number of Ju 88 C fighters.

One Ju 88 D-5 remains in good condition today in the U.S Air Force Museum in Dayton, Ohio. Three other Ju 88s are in crates at Farnborough, England.

The end: Soviet soldier with *panje* wagon in front of an abandoned Ju 88.

This Ju 88 A-4 was destroyed in a strafing attack on a Czechoslovakian airfield.

Captured Ju 88 A-5 on the airfield of the American 322nd Bomb Group in England.

Ju 88 A-5 long-range reconnaissance aircraft captured by the Americans and brought to the United States for testing. An in-flight photograph of this aircraft appears inside the back cover of this volume.

Technical Data

Type	Ju 88 A-1	A-5	A-4	A-15	B-0	C-4	C-6	D-1
Purpose	Bomber	Bomber	Bomber	Bomber	Aufklärer	Zerstörer	Nachtjäger	Fernerkunder
Crew	4	4	4	3	4	3	3	4
Power Plants	Jumo 211B-1	Jumo 211G-1	Jumo 211 J	Jumo 211 J	BMW 801 MA	Jumo 211 F	Jumo 211 J	Jumo 211 B
H.P.	2 x 1175	2 x 1200	2 x 1410	2 x 1410	2 x 1560	2 x 1420	2 x 1410	2 x 1175
Span m	18,25	20,08	20,08	20,08	20,08	20,08	20,08	20,08
Length m	14,35	14,45	14,36	14,36	14,45	14,96	14,96	14,36
Height m	5,30	5,07	5,07	5,07	4,45	5,07	5,07	5,07
Wing Area m2	52,5	54,7	54,7	54,7	54,7	54,7	54,7	54,7
Weight, Empty	7250	8050	8550	9000	9100	8000	8100	8480
Equipped kg	10360	12450	12-14000	12800	12470	11350	11450	11490
Weight, Loaded kg	2500	2500	2400-3600	3000	2500	-	-	-
Bomb Load kg	450	475	440	410	540	495	500	475
Maximum Speed kph	370	370	385	370	510	485	490	425
Cruising Speed kph	140	140	140	140	175	140	145	140
Landing Speed kph	9350	8500	8500	8250	9050	8600	8800	8600
Ceiling m	1500	2950	2500	2000	2850	3050	2950	2950
Range km	-	1150	1800	1800	750	870	875	875
Takeoff Distance m								
Equipment	FuG 16 FuG 25	wie A-1	wie A-1	wie A-1	wie A-1	wie A-1	SN 2	wie A-1
Armament	3-4 MG 15	5 MG 15	5 MG 81 evtl. 1 MG 131	2 MG 15	3 MG 81Z	1 MG/FF 3 MG 17 1 MG 15	3 MG 17 3 MG 151/20 1 MG 81Z	3 MG 15 2 Rb

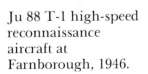

Ju 88 T-1 high-speed reconnaissance aircraft at Farnborough, 1946.

Technical Data

Type	Ju 88 D-5	R-2	G-1	G-6	G-7	H-1	P-4	T-1
Purpose	Fernerkunder	Nachtjäger	Nachtjäger	Nachtjäger	Nachtjäger	Fernerkunder	Schlachtflzg.	Fernerkunder
Crew	4	3	3	3	3	3	3	3
Power Plants	Jumo 211 J	BMW 801D-2	wie R-2	JUmo 213 A	Jumo 213 E	BMW 801D-2	Jumo 211J-2	BMW 801 G
H.P.	2 x 1410	2 x 1700	2 x 1700	2 x 1750	2 x 1750	2 x 1700	2 x 1420	2 x 1700
Span m	20,08	20,08	20,08	20,08	22,00	19,95	20,08	20,08
Length m	14,45	14,96	15,50	15,50	15,50	17,65	-	14,84
Height m	5,07	5,07	5,07	5,07	5,07	5,07	5,07	5,07
Wing Area m2	54,7	54,7	54,7	54,7	56,0	54,7	54,7	54,7
Weight, Empty	-	-	-	-	-	-	-	-
Equipped kg	11300	11500	12100	12400	13120	-	11400	12215
Weight, Loaded kg	-	-	-	-	-	-	-	-
Bomb Load kg	480	580	540	580	643	445	390	640
Maximum Speed kph	430	510	480	510	590	410	370	570
Cruising Speed kph	140	160	165	170	175	140	140	175
Landing Speed kph	8700	9200	9400	9550	9800	8500	8000	9800
Ceiling m	3100	3000	2800	2200	2220	5150	2000	1090
Range km	1150	750	720	700	1250	1300	1300	700
Takeoff Distance m								
Equipment	FuG 16,	FuG 220	FuG 220	FuG 220,227	FuG 220,228	FuG 200, 3 Rb	FuG 16,25	FuG 16,25
Armament	FuG 25	4 MG 151/20	6 MG 151/20	6 MG 151/20	teilw. 240	2 MG 131	2 MG 81Z	3 Rb
	z. T. FuG 200							
	3 Rb	1 MG 81Z	1 MG 131	1 MG 131	6 MG 151/20	WTZ 81		2 MG 131
	2 MG 17,				1 MG 131			
	2 MG 81							

Left: A Ju 88 A-5 of KG 30 abandoned in North Africa for lack of fuel.

Right: English soldiers inspect the remains of a Ju 88 shot down over France.

PANZER

A PICTORIAL DOCUMENTATION

HORST SCHEIBERT

SCHIFFER MILITARY

Stuka-Pilot
HANS ULRICH RUDEL

HIS LIFE STORY
IN WORDS AND PHOTOGRAPHS

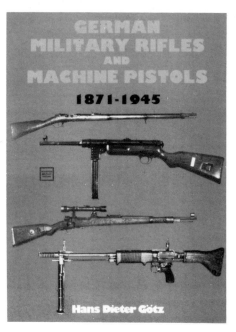

GERMAN MILITARY RIFLES AND MACHINE PISTOLS

1871-1945

Hans Dieter Götz

THE HG PANZER DIVISION

by
ALFRED OTTE

SCHIFFER MILITARY

• *Schiffer Military History* •

Specializing in the German Military of World War II

Also Available:

• The 1st SS Panzer Division - *Leibstandarte* • The 12th SS Panzer Division - *HJ* • The Panzerkorps *Grossdeutschland* •
• The Heavy Flak Guns 1933-1945 • German Motorcycles in World War II • Hetzer • V2 • Me 163 "Komet" •
• Me 262 • German Aircraft Carrier *Graf Zeppelin* • The Waffen-SS - A Pictorial History • Maus • Arado Ar 234 •
• The Tiger Family • The Panther Family • German Airships • Do 335 •
• German Uniforms of the 20th Century - Vol.1 The Panzer Uniforms, Vol. 2 The Uniforms of the Infantry •

THE GERMAN NAVY AT WAR 1935-1945 VOLUME 1
THE BATTLESHIPS
A HISTORY OF THE "BIG SHIPS"
•POCKET BATTLESHIPS•CRUISERS•
AIRCRAFT CARRIER "GRAF ZEPPELIN"

SIEGFRIED BREYER   GERHARD KOOP

THE GERMAN NAVY AT WAR 1935-1945 VOLUME 2
THE U-BOAT

SIEGFRIED BREYER   GERHARD KOOP

THE U-BOAT-LIGHT NAVAL WEAPONS-NAVAL INFANTRY DIVISIONS-NAVAL AVIATION-
HARBORS AND BASES-WARSHIP CONSTRUCTION YARDS-NAVAL CITATIONS-
UNIFORMS-INSIGNIA'S OF RANK AND SERVICE-FLAGS-

THE BATTLESHIP BISMARCK
A Documentary In Words and Pictures
ULRICH ELFRATH   BODO HERZOG